Ephraim Cutter

Is flour our proper food?

With some remarks upon the effect of animal food in consumption

Ephraim Cutter

Is flour our proper food?
With some remarks upon the effect of animal food in consumption

ISBN/EAN: 9783337201494

Printed in Europe, USA, Canada, Australia, Japan

Cover: Foto ©berggeist007 / pixelio.de

More available books at **www.hansebooks.com**

OUR PROPER FOOD?

WITH SOME REMARKS UPON THE EFFECT OF

ANIMAL FOOD IN CONSUMPTION,

BY

EPHRAIM CUTTER, A. M., M. D.,

OF CAMBRIDGE, MASS.

A LECTURE DELIVERED BEFORE THE N. H. STATE MEDICAL SOCIETY, JUNE 16, 1875.

CONCORD, N. H.:
PRINTED BY THE REPUBLICAN PRESS ASSOCIATION.
1875.

IS FLOUR OUR PROPER FOOD?

BY EPHRAIM CUTTER, M. D., OF CAMBRIDGE, MASS.

Gentlemen:

In asking this question, allow me the privilege of making the request that your society will at some future time answer this question, and officially communicate the result to the public, either through your own secretary, or through the speaker. I do so because I regard bodies like your own the only legitimate organizations competent to answer questions of public hygiene, and at the same time sufficiently authoritative enough to produce decisions which will command recognition and respect. A combined, unanimous, and officially announced opinion of the now existing New England Medical Societies, on any subject connected with social or sanitary science, would have more weight with the present generation than the testimony of all the classical medical writers of the past.

"It has lately been urged by Liebig that saline matter has failed to receive its due consideration as a nutritive element of food. It is perfectly true, as he has pointed out, that in the preparation of food for human consumption the natural article is often considerably depreciated in nutritive value by the abstraction that may happen to have occurred. Meat soaked or boiled in water loses more or less of its soluble portion, and included in this are its nutritive salts. Roasted meat on this account is of higher value than boiled. In the process of salting, a portion (about fifteen per cent., Liebig says) of the nutritive juice escapes into the brine. In the boiling of vegetables, nutritive principles, and particularly the nutritive salts, are removed by the water. The separation that is effected in the dressing of flour leaves this product in an inferior position to the grain from which it is derived. Both the saline and nitrogenous matters belonging to wheat are chiefly encountered in the outer or tegumentary part of the grain, and are, therefore, more or less excluded from white bread. It is a scientific fact, Liebig remarks, which Magendie has proved by experiment, that a dog dies if fed on white bread, while its health does not suffer at all if its food consist of brown bread, or bread made of unbolted flour. Liebig also asserts his belief that many millions more men could be daily fed in Germany, if it were only possible to persuade the population of the advantage which bread made of unbolted flour has over that ordinarily eaten." Pavy on Food, p. 145, 1875.

In this doctrine of Liebig, Pavy is not disposed to coincide, but thinks that in the mixture of animal and vegetable food the loss alluded to is supplied ; and that if, in eating bread made from white flour, we from our taste prefer to reject a portion of the wheat, "it does not follow that in so doing we are committing an act of prodigality, for what we do not use ourselves may be, and in reality is, turned to account in feeding animals that are either kept for some useful purpose, or reared for consumption as food ; and, in the latter case, the nutritive salts which we originally rejected in separating the bran from the flour may actually reach us amongst the constituents of animal food." Yes,—but do mankind, who live mostly upon flour, get this animal food in connection with flour ? I think not ; and I wish to express my dissent with Pavy on this point, and desire to be allowed to suggest *that the universal and exclusive use of flour, as found at the present time among the nations of Christendom, may (as Liebig suggests) result in disaster to the human race in the following particulars:*

I. May it not be possible that the use of flour is a cause of the prevalence of diseases of the nervous system?

II. May it not be possible that the use of flour is a cause of the present lamentable and astounding prevalence of late erupting and decayed teeth?

III. May it not be possible that the use of flour is one cause of the present prevalence of weak and diseased eyes?

IV. May it not be possible that the use of flour is one cause of the prevalence of baldness and premature gray hairs?

V. May it not be possible that the so-called change in the type of disease may in some measure be due to the use of flour, so universal for the past forty years?

VI. May it not be possible that the use of flour is one cause of the prevalence of some of our chronic diseases, as catarrh and consumption?

VII. May it not be possible that the use of flour is one cause of the numerical decline in our native population in New England?

We shall proceed briefly to give some of the reasons why we ask the privilege of making the above seven propositions, which, if proved to be true, affect and come home to every civilized human being, and in which all have a deep and vital interest as citizens and as Christians; for without sound minds and sound bodies states will decay and pass away, and the affections and intellect will be so warped and hindered that a full, symmetrical, and well-developed moral and religious character is an impossibility.

Dr. Kirkes, assisted by Dr. Paget, both of London, more than twenty years ago wrote as follows: "No substance can afford nutriment, even though it contain *all* the elements of organic bodies, unless it have all the natural peculiarities of organic composition, and contains incorporated with its other elements some of those derived from the mineral kingdom, which, as incidental elements, are found in the organized tissues, such as sulphur, iron, lime, magnesia, phosphorous, &c." Now, there is no other article of food that ordinarily receives at the hand of man such manipulation and processes of separation of mineral constituents as wheat, unless we except over-boiled food. If the wheat was subjected only to the simple process of one grinding, meal would be the result containing all the elements of the grain, which we may safely say that our Creator designed as the most suitable food for man. Used in this form, we believe that there would be much less decay and disease than at present exists under the use of flour, which is made from the wheat meal by the miller submitting it to bolting, sifting, or dressing. These processes form pollards and sharps and bran (as they say in England), or, in our vernacular, middlings, &c. The finer the flour is dressed, the whiter the bread, the less nutritious. Indeed, the result is much different from the original wheat,—how different, is shown by the following analysis:

From tables published in Johnson's "How Crops Grow," a most interesting and valuable book, deserving a place in every library in the land, we ascertain the following:

Composition in 1000 parts of substance of

	Water.	Ash.	Potash.	Soda.	Magnesia.	Lime.
Wheat grain,	143	17.7	5.5	0.6	2.2	0.6
Fine wheat flour,	136	4.1	1.5	0.1	0.5	0.1

	Phosphoric acid.	Sulphuric acid.	Silica.	Sulphur.
Wheat grain,	8.2	0.4	0.3	1.5
Wheat flour,	2.1	0.0	0.0	0.0

Amount of starch in wheat, ... 59.5 per cent.
" " " flour, ... 68.7 " "
Albumenoids in wheat, ... 13 " "
" " flour, ... 11.8 " "

It will be seen that there is a considerable withdrawal of mineral elements in flour, while the starch is about the same. The withdrawal of potash is $5.5 - 1.5 = \frac{40}{55}$, not quite $\frac{4}{5}$; of soda, $0.6 - 0.1 = \frac{5}{6}$; of magnesia, $2.2 - 0.5 = \frac{17}{22}$; of lime, $0.6 - 0.1 = \frac{5}{6}$, the same as of soda; of phosphoric acid, $8.2 - 2.1 = \frac{61}{82}$, almost $\frac{7}{8}$.

Now, as we have stated before, with the exception of over-boiled meats and vegetables, there is no one article of food that contains so small a quantity of saline ingredients, as shown by the ash, as flour, when compared with the normal amount in wheat. By our table it is seen that flour contains only about *one fourth* of the salts which nature intended man should get when he ate wheat. It should be stated that different specimens of wheat vary in the amount of their salts, as in their other constituents, according to the soils in which they have been raised. This variation lies between healthful limits, and is immensely less than the variation of flour, animal constituents, and that of wheat. Payen gives the amount of mineral matter in wheat from Venezuela at 30.2 in 1000 parts; do. from Africa, 27.1 in 1000 parts; Taganroy, 28.5 in 1000 parts; soft wheat from Brie at 27.5 in 1000 parts; soft from Tuzelle, 21.2 in 1000 parts; our table, at 17.7, gives a less amount than Payen. Taking 30.2 as the maximum, the deficiency of salts in flour would be between six sevenths and seven eighths, so that the amount of saline in flour is only about one seventh, instead of one fourth by the former table; or you may more correctly state the range from one seventh to one fourth, or, to put it the other way, FLOUR IS THAT ARTICLE OF FOOD IN WHICH THE MINERAL ASH IS DIMINISHED ABOUT THREE FOURTHS TO SEVEN EIGHTHS.

I. *May it not be possible that the use of flour is a cause of the prevalence of diseases of the nervous system?* In the treatment of these special applications of the subject, the reader may notice some repetitions. It is hoped that the importance of the subject will be a sufficient excuse, as the present effort is intended for practical value, and not as a specimen of choice English literature.

The Roman soldier, in the time of Julius Cæsar especially, was the type of the most vigorous manhood probably that the world ever saw. For fortitude and endurance in warfare, labor and suffering in campaigns, and perseverance under hardships, his reputation has not, upon the whole, been surpassed. In his day there were no railroads for transport, not many bridges for passing rivers, no Goodyear to supply india-rubber for protection to feet, head, or body. He had no pontoons, or telegraph, or balloons. No powder, or gun, or rifle, or cannon aided him in destroying his enemies. His was a hand-to-hand conflict with javelins, swords, and battering-rams. He interviewed his foe in person, and such was his individual physical power and development that his opponents almost invariably succumbed, and Cæsar was master of the known world.

How did the Roman soldier come to possess such a wonderful strength of physical and mental organization that he could accomplish feats of prowess which fill so large a space in the history of the world?

We know how he lived,—out-of-doors, inhaling plenty of pure oxygen, and exhaling carbonic acid, which was immediately born off, and its place supplied with fresh air. If he had lived in one of our modern houses, heated with stoves, and laboriously shut up air-tight, with no ventilation except an occasional opening of a door, breathing an atmosphere tainted with carbonic oxide and carbonic acid gases, besides the animal exhalations, window-blinds and sashes closed, and curtains drawn (which is the general average condition of New England houses of to-day), we think that his animal (*anima* means breath) powers would not have allowed him to accomplish his historic achievements.

But the Roman soldier, besides breathing, had to eat. No matter how much fresh air and exercise he had, his physique would have failed with imperfect food. He could not have developed muscle enough to climb mountains, swim rivers, fight hand-to-hand fights, and endure privations, unless his digestive organs had been fed with aliment which supplied the waste of tissue consequent upon exertion, and the withdrawal of the nerve force, vitality, or life, or whatever you are pleased to call that dynamic power which carries along the currents of our physical existence. In speaking of muscular actions, we are apt to regard the muscles themselves as the source of power. But if we should separate the nerves which connect the given muscle, or set of muscles, with the spinal or cranial system of nerve centres, it would be found that they would become as powerless for action as the engine when steam is cut off, or as inactive when its connecting belt with the motor is slipped off, so that it is more in accordance with the facts to speak of the nerve force as the primal power of all the muscular forces of the body. In this light we cannot conceive of the Roman soldier as other than a person of immense nerve power. It might not have been an intellectual nerve power, but it must have been a neurotic power sufficient to wonderfully sustain and control the still more wonderful combination of mechanical forces found in the muscular system. It is very generally acknowledged by physiologists that there is more or less waste of nerve and muscular tissues during the exercise of the varied functions of the human body. No light is seen, no sound is heard, no touch is felt, no smell is perceived, respiration is not kept up, digestion, secretion, excretion, cerebration, phonation, and muscular movements are not performed without a waste of the tissues which are the agents of the functions named. Now, the Roman soldier must have had just this tissue destruction, and he must have supplied it in his food, or else he would have broken down under such severe tests. History shows that he did not break down, and it is a very interesting question what he mainly subsisted upon. In looking over the list of the commissary department of the Roman army, we do not find the modern diet table. *Frumentum*, grain, or wheat was the main article of diet. A bag of wheat was a regular part of the outfit. It was whole wheat,—not flour. When the soldier was hungry, all he had to do was to eat it by chewing it whole on the march; or at a halt, or in camp, soaking it in water, and then, rubbing up with a stone, eat it either uncooked or boiled. Any of the animals he might chance to find were caught and appropriated as additional food, and were so much clear gain. There might be at stationary camps other articles of diet, but in the long run unbolted wheat was his principal food. This being the case, it is clear that open air life and wheat are, or were, the elements that can make a perfect physical organization. It is not here asserted that no other combination of fresh air and food does not furnish the same food data, but it is desired to emphasize that wheat has the undisputed character of a perfect food. Dr. Nichols, editor of the *Boston Journal of Chemistry*, says he entertains the profoundest respect for a grain of wheat: "It is a most marvellous combination of substances, admirably adapted for the building up and sustenance of the tissues of the human body." It is emphatically the food of mankind. Its history is traced back to the earliest ages. It has been found buried with the mummies of Egypt. Our modern civilization has adopted it, or, rather, preparations from it. The raising of wheat, and carrying it to market, occupy the attention of large portions of the human race. Last year (1873) it took two hundred and twenty-five full-sized ships to carry the surplus crop of California wheat to the markets of the world. The manipulation and consumption of flour from wheat furnish employment for a much larger number of people than the producers and freighters; and if we include those who eat .the food prepared from wheat and wheat flour, there is hardly an individual in any civilized community throughout the world who does not come into the most intimate relations with bread, pies, cake, puddings, gruels, crackers, muffins, dumplings, &c., all derived from wheat. The consumption of flour as food being so universal and large, may we not be allowed to infer that the characteristics of the tissues of the bodies of our race must be determined in some manner by this flour?—because these tissues are built up, nourished, and sustained by the food which is consumed.

The old Roman soldier was a perfect animal in organization; and may we not deem it reasonable to conclude that his diet may have made him, or that he could not have attained his condition without his wheat or some other analogous grain? What diseases were prevalent among his comrades we know not, as no hospital records have been handed down to us. In the face of what we know and have already stated, may it not be admitted that generally he possessed good health?—for no sick or diseased soldier could have done the work that was accomplished.

Now, of what did his wheat food consist? As we have already seen, it had a mineral ash, varying from 30.2—17.7 in 1000 of substance. That it had phosphoric acid, 8.2; potash, 5.5; magnesia, 2.2; lime, 0.6; sulphuric acid, 0.4; silica, 0.3; sulphur, 1.5.

Note that phosphorus or phosphoric acid is found largely in the albumen of the nervous tissues. It is also found in the bony tissues.

Chemical constitution of nerve (Vangeala):

Albumen,	7.00
Fat,	5.23
Phosphorus,	1.50
Osmazome,	1.12
Acids, salts, sulphur,	5.15
Water,	80.00
	100.00

Albumen is found solid in nerves. Its composition, according to Scherer, is as follows:

Carbon,	54.9
Hydrogen,	7.0

Nitrogen,	.	. 15.0
Oxygen, Sulphur. Phosphoric acid, }		22.4

But our table shows that there is no sulphur, sulphuric acid, or silica in flour; that it contains only 4.1 parts of ash, 2.1 of phosphorus, 0.6 each of lime and soda, 0.5 of magnesia, and 1.5 potash, the quantities varying from one fourth ash to one eighth of what the soldier in question fed upon.

Here there is a withdrawal in flour of nearly seven eighths of the proper nerve food, phosphorus, found in the wheat, the main ration of the old Roman soldier. It is probably the soluble and assimilable form of phosphorus,—one that the digestive system can absorb, and the nutritive system appropriate to its sustenance. Thus, modern, civilized mankind are generally living upon a food which is deprived of seven-eighths of its nerve-producing, sustaining, and corroborating element, *phosphoric acid*.

We raise the question, seriously, which stands at the head of this section. Does the use of flour promote (that is, assist, predispose to) affections of the nervous system? Mark, we do not ask whether it is the sole existing cause, but whether, if mankind now received in its bread eight eighths of phosphoric acid instead of one eighth, there would not probably be less disease of the nervous system.

Eight eighths were designed for man's use by the Creator. Eight eighths gave the Roman soldier his nerve energy and muscle. Suppose he had had only flour bread, and got one eighth: would he not have sensibly suffered? Could he have carried his sixty pounds of baggage? Indeed, we find that the absence of what Cæsar calls *frumenta*, corn, or grain (not our maize, or Indian corn, which was then undiscovered), *par excellence*, or wheat, from their rations, was the cause of tumults, disturbances, and sometimes war. Suppose Cæsar had started a first-class modern flour mill, and, separating almost seven eighths of the nerve food from their wheat, had fed his soldiers with the unnatural manufacture: may not we be allowed to think there would have been equal trouble?—for one cannot imagine such a large diminution without a corresponding lack of tonicity in those tissues needing and accustomed to a full supply. To put it differently, suppose Cæsar had removed 87½ per cent. of his soldiers' proper nerve food from their wheat: would he not have had a right to expect only 12½ per cent. of energy, tone, or vital force in those soldiers' nerves? And yet, this is just the state of things our boasted modern civilization has put us into. Because public opinion says that the whiter and lighter the bread is baked the better it is, therefore all Christendom acknowledges the declaration, and eats the food, which contains the less of solid, substantial elements, the whiter and lighter it is. It cannot be denied that neurotic complaints are very common and chronic. Never were there so many insane people; never were physicians called upon oftener to treat nervous diseases than at present. How often people drop dead from heart disease, found upon examination to be solely from the want of proper inervation. How marked is the prevalency of paralysis. How the nerves of special sense suffer. We have trouble with the eyes, very commonly. Our children, if we have any, grow up thin, ethereal, *nervous*, anæmic. They die of consumption, and break down readily under the discipline of schools. Then see what a vast amount of nervous diseases in woman, in every condition and class of society. Go into any public assembly in New England; see the cry of distress and care impressed upon the countenances,—a cry for something they lack. It is a beseeching look. Some say it is from hard work. Well, it is hard work to fight the battles of life with but 12½ per cent. nerve food.

May it not be that the diet of our farmers, *white bread, pies, cakes, doughnuts, crackers*, deficient as they are in the full amount of nerve food, is partly the cause of their own, and particularly their wives' decay, and distressed looks and decayed teeth, and weak nerves that tremble and shake and

ache when engaged in services which should be pleasurable,
—not painful? Consider, also, the amount of nerve force it
takes to digest the starch, which is a main constituent of
flour, compared with the amount required to digest animal
food containing the same amount of nerve food. Sometimes cases of dyspepsia (difficult digestion) seem to depend
upon the fact that the nerve power (so scantily fed upon
flour) is all used up in labor and work, and in carrying on
the other functions of the body, so that there is none left to
digest the food. In other words, the system is too tired to
eat. What follows? As a matter of course, the whole system is unnourished, the other functions fail in their full performances, and if this be carried too far the nervous system
rebels, and we have neuralgia, headaches, and distress in
various parts of the body ; and if these things be continued,
disease results, sometimes followed by death.

The fact is, that we are surrounded constantly by the
causes of disease. Vegetation is subject to the same law.
The moment animal and vegetable systems are reduced in
their vitality, then step in parasites, animal and vegetable,
which are called disease. In potatoes, for instance, that rot,
it has been found that there is a withdrawal of lime to nearly
75 per cent. of the normal quantity. The aphides and fungi
and microscopic algæ prey upon the tubers, and by some
are thought to be the cause of the potato rot ; but as they
are found wherever there is decay, animal or vegetable, it is
more probable that the loss of the mineral constituents so
weakened the vitality of the potatoes that they fell an easy
prey to the insects and spores which are everywhere present,
ready to act if they get a chance. Our present system of
agriculture allows the ground no time to rest, and when the
soluble salts of mineral potash food are exhausted from the
soil, plants grown in that soil cannot get their proper mineral
food. The old Mosaic law, of letting the ground enjoy the
rest of a sabbatic year (once in seven), allowed the undissolved
lime, soda, potash, magnesia, salts, &c., to become soluble

under the atmospheric influences, so that when the land came to be planted the next year, it possessed the materials in a soluble form for making growths with their full amount of mineral constituents. They built up, the plants resist the aphides and the fungi, and, as people say, they do not rot.

Dr. Nichols, the able editor of the *Boston Journal of Chemistry*, said that when he gave his land a dressing of salts (sulphate of magnesia) then it bore perfect wheat, while before it was a failure. The same gentleman has a cold-grapery in which he raises large quantities of perfect and most beautiful fruit, entirely free from rust, mildew, smut, mould, or insects. The peculiarity of this grapery consists in having no manure but mineral manure, in the form of salts of the various alkaline earths. The supply was put in the border outside, and is calculated to last for thirty years. Eight of these years have passed, and the abundance, perfection, and beauty of the fruits are a growing and indisputable comment upon the Doctor's wisdom, and the law of the indispensableness of mineral food to perfection in vegetation.

If mineral salts are so necessary to healthy vegetation, is it unfair to reason that animal life needs them just as much? And as nerve force is so indispensable a part of animal life, do we reason incorrectly when we assert that in our opinion nervous disease would not be so prevalent if the human system were fed with all the 100 per cent. of phosphoric acid that God intended it should have? Ours is such a bustling, active, nervous age, that we need more nerve food than ever before in the history of the world. How many of us wear out, how many of us suffer, how many of us fail from want of proper nerve food, none can tell. One thing is certain: the old Roman soldier did not give out until the introduction of wealth brought on an age of the most extravagant living the world ever saw. If the diet and habits had been kept down to the wheat standard in the palmy days of the empire, Rome, too, might have withstood decay (other things being equal) a much longer time. And what perpetuity can we expect for our own country if we rear a weak race, with feeble, nervous systems, on food which has lost nearly 87½ per cent. of phosphoric acid?

The subject is one of importance. If the case has been made out even feebly, public opinion should be moulded in the right direction. This can be done only by united, continuous effort of right-minded people. Let these, when convinced, say so, and become centres around which opinion may crystalize.*

II. *May it not be possible that the use of flour is one cause of the prevalence of late erupting and early decaying teeth?* Dr. Clough, of Woburn, Mass., in his lecture upon "Teeth," says, unequivocally, yes ; so, also, does his nephew, Dr. Harriman, of Tremont Temple, Boston, in some very able articles on the decay of teeth, say yes ; and so do other dentists ;— but the public are not aware of it, and probably never will be, until this conviction is branded in by lecturer after lecturer, dentist after dentist, physician after physician, or, rather, medical society after medical society ; for there is nothing so difficult as to inform, affect, mould, or create thus a healthy public sentiment about what we shall eat or wear. Allow me to introduce a short paper upon this subject, published by myself in the *Boston Journal of Chemistry*, December, 1874, giving the writer's sentiments.

There is no doubt that the decay of teeth prevails to an alarming extent, and it is very humiliating to our modern civilization to have it characterized so generally by the occurrence of diseased teeth. The ætiology of this disease is a great, broad, and deep subject. No doubt, many elements combine to cause it ; and the person who should positively announce a single agent would be dismissed as unworthy of attention. Still, it is a matter worth discussing, and deserving the attention of the ablest minds. To ignore is not to arrest ; hence we offer a few suggestions for consideration.

"The *Chemical News* ascribes the potato rot to a defi-

---

* *Boston Journal of Chemistry*, February, 1875.

ciency of lime and magnesia in the soil. Different observers state the percentage of magnesia in the ash of sound tubers at from 5 to 10 per cent.; in the diseased tubers an analysis shows only 3.94 per cent. Analysis of sound tubers shows over 5 per cent. of lime, but in the diseased tubers only 1.77 per cent. was found. A similar observation was made some years ago by Professor Thorpe, with regard to diseased and healthy orange-trees. In the former there was a deficiency of lime and magnesia."

According to these authorities, a deficiency of mineral salts in the vegetations named is supposed to be a sufficient cause for decay. Now, it is an interesting question whether there is any article of food employed by mankind which is deficient in mineral matter. If so, then it should be made known to every family in the land.

Perhaps there is no article of food more generally consumed than flour, *i. e.*, wheat flour. In the forms of bread, cakes, and pastry of all kinds, it enters into every house, and is universally used and regarded as the "staff of life."

Does flour possess a requisite amount of mineral matter?

To answer this question, Mr. Sharples, the well-known chemist, analyzed for me the "Peerless Flour." He found 0.55 per. cent. of mineral ash, a little over half of one per cent. He stated, also, that the proportion of ash in the whole grain varied from 1.65 to 2.50 per cent., so that the diminution of mineral food varies from two thirds to four fifths. In other words, *by the use of flour, mankind loses from two thirds to four fifths of the elements that go to make up teeth and bony structures.* This statement deserves to be written in letters of gold over the door of every bakery and kitchen in the land. Flour has been used for generations, and, if we can rely upon Mr. Sharples's statement, mankind has all this while been deprived of the greater moiety of the mineral food that the Almighty intended it should have the benefit of. Is it not natural to expect that the bony structures should suffer from this great withdrawal?—for it is a *great* withdrawal. Suppose that a water supply-pipe should be cut off two thirds to four fifths: would not the supply be greatly diminished? Cut off the same quantity of time from the hours of daylight: would not our darkness be great? Take away two thirds to four fifths of our food: would not a strong man become very weak? Cut off air to the same extent: would we not lose our breath? And why should the bony tissues not suffer in like manner when their food is withdrawn? I think they do. Perhaps a little evidence in the contrary direction may throw light upon this.

A dentist, whose name is well known, said that he filled some fourteen cavities in the teeth of his first-born child by the time he was four years of age. He put his family upon the use of the whole grains, and the next child had no retarded dentition, and not a decayed tooth up to the same age. The same gentleman says that the teeth which decay are not compacted or knit together with the firmness of healthy teeth. There seems to be an arrest of perfect development. Though what can be more natural than to repeat imperfect development and decay, when from two thirds to four fifths of the proper bone food is habitually withdrawn from an article of diet which is more largely used than any other?

The Hon. Mr. Smith, U. S. Commissioner of Indian Affairs, at Washington, D. C., said, when asked if he had ever seen Indians with decayed teeth,—" Rarely." His experience with Indians is very intimate, and covers a period of a quarter of a century. They live on animal food.

How common it is to see infants not cutting any teeth at all, until they are twice as old as they ought to be. The good effect of the whole grain diet is shown in Dr. Harriman's second child above alluded to. To be sure, it is only *one* case, and must not be made too much of.

Now, what is to be done about it? Certainly one man's dicta amounts to but very little alone. What we need is evidence from others. Suppose that every medical association of Massachusetts take up this matter, have analyses of flour made, try feeding mothers and children upon the whole grains of wheat meal, oat meal, corn meal, beans, &c., and suppose they all come to the same conclusion as the writer has done; suppose they officially announce the result: would it be long before the general public would heed the truth, and thousands of persons would rejoice in the possession of that priceless treasure, a set of perfect teeth?

NOTE. Through the kindness of Mr. E. H. Davis, Superintendent of Public Schools in Woburn, Mass., the writer has been furnished with the following *astounding* statistics, embracing returns from several of the largest primary schools of Woburn, a fair representation of the prevalence of diseased teeth among children:

	No. scholars.	With sound teeth.	Decayed.
Lawrence primary,	113	13	100
Plympton street primary,	94	27	67
Highland street primary,	71	25	46*

III. *May it not be possible that the use of flour is one cause of the prevalence of weak and diseased eyes?* Dr. Hasket Derby, of Boston, the eminent oculist, in a late conversation with the writer, said those conditions which caused or promoted the decay or weakness of teeth, also would cause or promote diseased conditions of the eyes. In fact, he stated that he largely prescribed phosphates in order to supply this element to the ocular tissues. Dr. Budd calls attention to the prevalence of ulcerated eyes (corneas) among the Hindoos that feed exclusively on rice. Now, the ash of rice is only 0.90 of one per cent. (Payen), and that of flour 0.41, according to Johnson. Is it not fair to conclude that if rice will not give vitality enough to secure perfect eyes, flour, which has an ash less than one half the ash of rice, will not confer a perfect vitality also, especially when it is relied on

---

* See *Report State Board of Health, Massachusetts*, 1875. Of 880 of the school children in Woburn, Lexington, and Bedford, in 1874, under twelve years of age, *two thirds* had decayed teeth.

so much at the present time as the staple article of food for growing persons?

In this connection we would allude to the fact that ulceration of the cornea occurred in dogs kept exclusively upon sugar. It is well known that human beings, who live upon scanty and impoverished diet, and who are exposed to vicissitudes, present a larger number of eye affections than those from the higher and better fed walks of life. When the writer, in his medical pupilage, attended the Blockley Hospital at Philadelphia, containing 3,000 inmates, the eye patients used to be ranged for inspection in rows of about one hundred persons at one time. The spectacle was one that was startling, if not amusing. So that it is beyond a doubt certain that impoverished diet and diseased eyes are connected together; and certainly, if the withdrawal of three fourths to seven eighths of the mineral ingredients of an article of food is *not* an impoverishment, it would be difficult to find one anywhere.

In the report of the Massachusetts State Board of Health for 1874, Dr. Winsor gives the results of some inquiries made as to the prevalence of defective eyes among school children. Out of two hundred and forty pupils, fifty-five were found with an imperfect vision. We should think that 23 per cent. was a large number to be found among a healthy people. It is evidently much less than the prevalence of diseased teeth among the same class, to which we have already referred.

Out of 1,000 children under the age of eighteen years, in a large school in New York, 703 were found with defective eyes when examined with the opthalmoscope.[*]

We call upon the oculist to give this subject the attention it deserves. We also think it would be a work germain to the state boards of health to institute comparative inquiries as to the presence of defective eyes among flour-eating and non-flour-eating communities. We think the evidence ad-

---

[*] Dr. C. R. Agnew, *Meeting Am. Social Science Ass., Detroit*, 1875.

duced certainly does point in the direction that a substance, used as food, which is so defective in mineral ingredients as flour, may possibly act as one cause of anomalous and defective conditions of the organs of vision.

IV. *May it not be possible that the use of flour is one cause of premature whitening of the hair, and baldness ?* The hair appears to have a life of its own, like the teeth and nails. It is remarkable for the resistance it offers to ordinary decay, having been found in good preservation on mummies; and how many of us cherish locks of the hair of our dear departed ones, as imperishable mementos of their history. The hair has long been regarded as a sign of vigor, a notable case of which is seen in the history of Samson. Hence the falling of the hair in old age (calvities) is deemed a natural consequence of the impairment of nutrition, and the failing powers of life. Sometimes it is general, arising from causes that lower the vital tone, and withdraws the nutritive pabulum, as in syphilis, fevers, anæmia, rheumatism, gout, neuralgia, fast living, great study, violent emotions, dyspepsia, want of cleanliness, over purgating, and lastly, hereditary peculiarity, pregnancy, and *deficiency of formative forces inherent in the system.*

This is a portion of the list of causes laid down by Dr. Tilbury Fox, a late and eminent London authority in skin diseases.

The chemical composition of hairs is not well understood, but it appears that they are chiefly composed of a nitrogenous substance, soluble in alkalies, with the evolution of ammonia, and insoluble in boiling acetic acid. Some authors consider it a combination of protein with sulphur (Kolliker). Mulder considers hairs to be mainly a protein compound, with sulphamid (*i. e.*, N. H. $_2$S.), of which he finds ten per cent. The ash amounts to one to two per cent. It contains iron, manganese, silica, sulphur, magnesia, and alumina. Beyond a doubt, sulphur is a large and essential ingredient of hair. It is sulphur that gives the characteristic odor of burning hair. If this were withdrawn from the aliment entirely, we should expect to find imperfectly developed and easily falling hair. Grayness is considered the primal stage of the decay in hair. Hairs are constantly falling out, and as constantly renewed. Sometimes there are periodical castings of hair, as in the cases of horses shedding their coats in spring. Just so we have periodical castings of leaves from trees, as well as constant shedding in the evergreens. We have seen the effect the feeding of bullocks upon a diet without salt has upon the hair, making it starring, roughened, disordered, and the skin hide-bound; and, on the other hand, bullocks, fed with the same diet plus the salt, will be found with a smooth, glossy skin, showing, evidently, the effect of a mineral ingredient of food upon the hair, and demonstrating the intimate connection between it and the hair.*

Now, in repeatedly looking over the analysis of the ash of flour, the writer was impressed after a while with the significant fact that there was no sulphur in the flour, and having for a long time been painfully aware of the prevalence of premature baldness, without being able to find any very satisfactory cause assigned, we would make the query at the head of this section,—May not the use of flour promote baldness, and grayness of the hair? If it is a true proposition that the organized substances must have food that contains all the mineral ingredients of those substances, then, as flour has no sulphur, it is pretty conclusive that it is no suitable food for hair, which has ten per cent. of sulphamid, and that those who subsist mainly upon flour must expect imperfect hair, and premature grayness and baldness, unless they get their supply of sulphur from some other source.

Commissioner Smith, before alluded to, when asked if he ever saw a bald-headed Indian, said, promptly and unequivocally,—" Never."

---

* During the spring-time, neat cattle require double the amount of salt than at other times; and it is natural it should be so, as this is the season of hair shedding, and consequently there is more demand for saline ingredients to build up the substance; and as chemically pure salt is nowhere sold as an article of commerce, it is probable the animals obtain many other salines besides the chloride of sodium.

Another thing: it is generally a well received notion among physicians, based upon the researches of Gruby mainly, that the unnatural falling and bleaching of the hair is due to microscopic fungi in the interior of the hair itself (Herpes tonsurans), or under the epidermis and around it (Porrigo decalvens). Now, if the rot of the potato is found to be due to the absence of mineral salts, why not the decay of the hair, which approaches a vegetable appendage more than almost any other substance in the body? It would be interesting to have some qualitative analyses of hair made, to ascertain whether there is this deficiency of normal salts of sulphur. It is true, that grief, excessive intellectual activity, nervous influences, and severe demands upon the nervous system, are sometimes evidently concerned in these diseased conditions of the hair. Of course, they are outside of the present explication. One thing is very certain—the unanimous testimony of those who have visited the abodes of savage races is, that they are remarkably supplied with capillary covering of their heads, except when removed by accident or design. [Messrs. Baldwin & Botume, Jr., 122 North Market street, Boston, have packed pork for the last thirty years. In times past they have received hogs fed on the sweepings of flour mills, but the meat was so bleached, and the fat so readily fissuring up into distinct masses, that the pork was unmerchantable; and if all the hogs they packed were fed on flour, their business would be destroyed. They state further, that the bristles (hair) of the flour-fed hogs were very *white and thin*, so as to be worthless for the brush-makers.] In the absence of statistics as to the prevalence of baldness, we must give only general impressions.

At Gilmore's World's Peace Jubilee, held in Boston some years ago, there were present at least twenty thousand people at one of the afternoon concerts. The writer occupied a position in the rear of the audience, and as the rays of the sun illuminated the auditorium, he was very much impressed with the number of bald heads of the men present. It seemed as if nearly two thirds of the heads were gleams of shining light. It would be interesting if we could collect figures as to the baldness during the brown-bread era of New England. Some, who lived in those days, when questioned, testify that as far as they can remember there were less bald people than now. This suggestion is new, and needs further discussion and investigation. It is thrown out simply as such, in the hope that it may lead to some light being thrown upon the pathology and treatment of hair. It is safe to infer, from the above statement, that the chances for healthy hair are better with food that contains the full amount of mineral matter than with flour. Indeed, with Dr. Fox's specifically stating that premature baldness is directly caused by a deficiency of formative force inherent in the system, we may reiterate our statement by a syllogism:—The system is liable to baldness because it lacks formative force. The formative force of the system is derived from food. Flour is food lacking seventy-five per cent. of formative force. Hence the system fed upon flour (other things being equal) is liable to baldness.

V. *May it not be possible that the use of flour is one cause of the change in the type of disease, from strong (sthenic) to weak (asthenic)?* That the type of disease has changed, is a fact acknowledged by the regular profession of medicine; and this change consists in diseases assuming that prostrating type and character which require stimulating and supporting treatment of every kind. Patients with certain diseases formerly required, bore, and yielded to reducing treatment, such as general depletion and active purgation ; but now, when affected similarly, they are found not to bear the same measures without serious depression of the vital forces. Occasionally we meet with patients who do bear the old treatment, but they are of such apparent robustness and vigor as to make the indication very plain ; and it is possible that the profession have swung too far to the other extreme.

But the object here is to discuss the relations of a diet of flour to a weakened condition of system, which renders diseases of a low, inactive character more readily prevalent. Allow us to make a suggestion. Suppose, for instance, our farmers should manure their crops for one year with only one quarter of the fertilizers ordinarily deemed necessary; suppose they should do this, not for one year only, but for forty years: what sort of vegetation or crops would they have? Why, it is clear that the vegetation, as a general thing, would be stinted, the crops small and ill-developed, the vitality of the plants feeble; and instead of pruning, reducing, and cutting down, measures of a "supporting" character would have to be adopted, as it is doubtful whether the plants would stand up of themselves. Something cannot be made out of nothing—*ex nihilo nihil fit*—is an ancient motto. Now, apply this principle to mankind: let them feed upon food (as they have done) for forty years or more, which contains one fourth of the mineral food which they consumed in the days of coarse bread, made out of whole wheat, corn, rye, barley, and oatmeal: would you not expect a change in the character of the tissues and fluids of their bodies? As plants would show degeneration, feebleness, and stinted development, would you not expect the same result in animals? The doctrine of the indispensableness of mineral salts to vegetable growths is of modern origin. We wish to go further, and apply it to animal tissues likewise; and we feel assured that when the days of fine flour are past there will be the old-fashioned type of disease back again, because the bodies of the people will obtain their full modicum of mineral salts. Indeed, in remote rural districts, where the old ways of living are practised, the necessity for depleting measures is said to be as strong as ever.

Take our typhoid fevers, our scarlet fevers, diphtherias, measles, small-pox, choleras, dysenteries, cholera-infantums, and consumptions: nearly all get iron, quinine, beef-tea, wine, and stimulants,—and rightly, too, in most cases. Some of these complaints are thought by the best minds in the medical profession to be of a vegetable, parasitic nature, communicated to man from decaying vegetable matter. It looks now, more than ever, as if this is so. Should it be proved to be so without a doubt, then, as these smuts, rusts, ergots, mildew, etc., prey upon the vegetables of impaired vitality, so it will be seen that mankind, weakened by living upon food containing but twenty-five per cent. of the normal amount of mineral salts, has fallen an easy prey to diseases of a low type, and typhoid and parasitic character. This suggestion of the connection of the use of flour with the present asthenic type of disease is entirely our own, and no one but ourselves is responsible for it.

And here we would not press the parallel too far, as people commonly eat animal and other vegetable food than flour. We admit it; but how small a proportion of animal food enters into the diet of common people! As Dr. Salisbury says, we are two thirds animal and one third vegetable in our constitution. Yet our food is the other way: we eat about seven eighths of vegetable food, and one eighth animal. It is safe to assume that among civilized people to-day, the diet is mainly bread, and that bread is flour bread; so that the use of flour as food is so large as not to invalidate the general truth of our statement. It is fortunate we do have other food, for if confined to flour bread alone we should probably become so enfeebled as to die, like Magendie's dog, in forty days. It is desired to suggest whether a diet of flour, if not complete, would not cause such a degree of weakened tissues and fluids as to perceptibly diminish the power to resist disease, so that in the conflict with it flour-eating patients are found to be the " under dog in the fight," needing tonics, stimulants, and supporting treatment, to come out the " upper dog."*

VI. *May it not be possible that the use of flour is one element in the causation of chronic disease?* Among these we class skin diseases, cancer, tubercle, catarrh, neurotic disease,

---

* See *Boston Journal of Chemistry*, Feb., 1875.

tumors, paralysis, diseases depending upon the fatty degeneration, such as apoplexy, Bright's disease, &c. Now, these are diseases of nutrition mostly. The functions of nutrition and growth are not preserved in the normal state. We are not prepared to say but that diseases must and will come, but we do ask to suggest that there is no surer way of inviting diseases than to impair the vital forces of the tissues and fluids of the body by a diet from which seventy-five per cent. of the ash is withdrawn. There are hereditary, physiological, and local causes that are to be taken into account, and diseased conditions seem to have a number of causes, all coöperating and combined together in a chain to produce them. Break one link, and the whole chain is broken. The disease may not occur, because, so to speak, the keystone of the arch is absent. We wish to record our opinion that flour-eating is a very important link in the chain.

Then, as the chemical composition of starch is so much like that of fat,—$C_{12} H_{10} O_{10}$,—when the starch is in excess, would it not be natural to expect that the tissues would more readily turn into fat?

VII. *May it not be possible that the use of flour is one influence in causing the decline of our native population?* Calling in once more the simile of the plant fed for forty years in succession on one quarter of its proper ration of mineral food, it is seen that the reproductive faculties are stinted and narrowed. The crop of corn, it may be, is diminished ; the yield of potatoes curtailed to narrow amounts, and the potato population declines. No one, knowing the cause, wonders why. To have the reproductive organs, in the plant, the bud, the flower, and the fruit, in abundance and vigor, plant-food must be furnished in sufficient amount. But by the terms of our simile, seventy-five per cent. of this amount has been withdrawn. The native population of New England is to-day somewhat in the condition of the barren tree in the parable, which the lord of the vineyard was going to cut down because it cumbered the ground by bearing no fruit; but the gardener stands by, and says, "Not so, lord; don't cut it down till I dig around the tree, and dung it for another season, and then if it does not bear fruit it may be cut down." In other words, the gardener would stir up the soil, and break it up, so as to expose as large an amount of surface to the action of the atmosphere, and apply manure, —that is, soluble mineral food,—and then he would see if the tree, getting its full dose of vital force through this soluble mineral food, would not exercise its highest and culminating function in bearing fruit to please the eye, and afford nourishment to its owner. If not, why then it should be removed as cumbersome. Why should not animals have the same treatment and fate? "Herein is the Father glorified, if ye bear much fruit!" How can there be fruitfulness without the proper supply of mineral food? And if no children are born, how long will it be before the race becomes extinct, and cumbers no more the ground? Ask Dr. Allen whether families were small in the days of eating bread made of the meal of any of the cereal grains. Ask him, also, whether, in those days, there were two thirds of the children that were born having decayed teeth before they were ten years of age; whether the hair became gray at the age of forty years, or baldness common at an earlier age; whether so many young people used to wear spectacles as now; whether there were so many diseases of low type;—and then ask him whether the present low rate of reproduction may not be in some degree explained by referring the cause to a condition of withdrawal of nutritive material from their food, which no husbandman, or Arlington gardener, would allow to be practised upon the mineral food of his fruit-trees or market vegetables,—that is, give them only one quarter of the amount of manure ordinarily deemed advisable or expedient.*

Effect of diet in consumption. The effects of animal food

---

* In Bolles's Switzerland of America, according to Prof. Sharpless, when the Piute Indians are fed on flour and sugar, their reproduction is reduced almost to nothing.

in consumption (phthisis pulmonatis) is very remarkable and favorable. Under its use the cough will generally become less, the expectoration diminish, the night sweat cease, even if there are physical signs of diseased condition throughout one lung. When we understand that nitrogenous food is directly applied to the growth and renovation of the structures of the body, and that consumption, as its name signifies, implies a wasting of tissues and a destruction of the same, it is easily seen how appropriately animal food is adapted to the wants of the consumptive. The intimate nature of this disease is not thoroughly understood, or settled, even after so many years of investigation by thousands of studious physicians. Some attribute it (outside of hereditary predisposition) to defective alimentation, digestion, and nutrition, or, in other words, to the want of proper food, attended by any influence that depresses the vital powers. Recent experiments in Greece, France, and Germany have demonstrated that it is communicable,—in other words, contageous,—by contact. In this country, a physician, whose name I am not permitted to use, states that he has the records of one hundred post-mortem examinations of pigs dying of consumption, induced by his feeding them upon sour swill. The writer inoculated six rabbits with tuberculous matter about one year ago. None of the animals perished from consumption, although one had a curious tumor at the side of the inoculation, which continued for six months. It is probable that the perfect condition of the digestive organs resisted the disease. Now, without looking further into the causes of consumption other than to say, that, as a disease of debility and nutrition, it would be liable to occur among those who eat starch with a diminished amount of mineral food. We would like to relate, briefly, some few cases that have come under our own observation, premising that they have also occurred in other physicians' hands.

CASE I. Man aged twenty-one, seen in August, 1874, presented the physical signs of advanced tuberculous disease in both lungs; diminished resonance on percussion, coarse and fine expiratory rales, increased vocal resonance, no cavernous respiration. He had night sweats, a loss of appetite, flesh, and strength, continuous, harrassing cough with copious expectoration, pallid looks, with every expectation of going down to an early grave. Upon being put upon an exclusively animal diet (broiled beefsteak mainly), with a simple medicinal auxiliary treatment, his symptoms became ameliorated and improved,—so much, that in the space of about six months he considers himself almost entirely well. He has gained flesh and strength; the night sweats ceased long ago; the physical signs had, at the last time I saw him, almost entirely disappeared. In fact, he is so well that he has dug out the bottom of a well, and gone to work at daily labor in a shoe-shop. This young man had been living, before his sickness, almost exclusively upon flour bread. The animal food was very repulsive to him, and it was only by much urging and coaxing that he was persuaded to continue in it. Now, however, he is very much attached to the diet, and would not give it up.*

CASE II. The father of this young man was seen at the same time, sick with the same disease, longer standing, but the physical signs were not so marked in character as the son's; and I thought him a more favorable case, considering his age. He went upon the diet, with appropriate treatment, but died in a few weeks, saying that he felt very much better under the treatment, but that it was too late.

CASE III. A man aged about sixty-five, a very active currier, was seen in August, 1873. He was in great alarm about a recent hemorrhage from the lungs. There was cough, paleness, loss of flesh and strength; the heart sounds were abnormal. There was dulness on percussion, and expiratory crackling over the left upper third front; night sweats, loss of appetite. Moping spiritlessly about, he acted, if he not did speak, his sentiments of the grave character of

---

* He has given up, partially, the diet, and has relapsed, and will probably go on to die.

this disease. Animal food, exclusively, with a change of air, local medication, with tonics, in the course of the year wrought such a change in him that he resumed his business, and has continued in it up to the present time, although he has exposed himself rather unwisely. Now, it seems as if the full supply of nitrogenous food acted in combination with other means to restore the tissues to their normal power of resisting disease. The physical signs of diseased lung disappeared on the last examination. The cardiac trouble remained.

CASE IV. A man about fifty, in February, 1874, had a severe cough, wasting, night sweats, loss of appetite and strength, copious expectoration, diminished resonance, and expiratory crackling over the left upper third front. He was alarmed and feeble. Though distasteful, he pursued the animal diet, taking it as a medicine, and also took some alterative drug. The improvement in this case was marked and prompt. The physical signs, cough and expectoration, diminished and disappeared, and now he appears restored to a perfect state of health. It is proper to state that there might have been an error of diagnosis on my part; still, the case was one which, if not already tuberculous, was in a fair way to become one.

CASE V. Lady about thirty-five—advanced stage of disease, which had existed for three years. The symptoms of this case were ameliorated for a time by a rigid adhesion to beefsteak diet, and medical treatment addressed to throat, skin, and stomach, but without success; for, rather unexpectedly, she grew worse, and died in the city of New York. The digestion improved, and the diet was not distasteful; but the disease was too far advanced to be checked.

CASE VI. A woman, aged about fifty, was seen in July, 1872. She had buried a daughter the season before in consumption, and now she was attacked with the symptoms of the same disease (was it communicated by contagion?),—the emaciation, the cough, the expectoration, the physical signs of dulness in the left lung, and crackling expiratory respiration were well marked. Under the animal food and appropriate treatment she regained her health, and is living to-day in the enjoyment of it.

CASE VII. Man aged fifty; sick a year; pale, emaciated, coughing hard, cavernous respiration in the left upper third front of the left lung, night sweats, chronic dyspepsia for half his life; no appetite for anything; great aggravation of the gastric symptoms. The effect of the animal diet was to revive his appetite very strongly for those things he should not eat. The cough, expectoration, and nutrition improved; but so strong and irresistible was the desire for forbidden food, that he gave up his restrictions, went on to eat anything he liked, and died in the course of two months.

CASE VIII. A woman thirty-four years of age,—father and mother both dead of consumption,—with decided marks and physical signs of developed disease in both lungs, was placed upon animal food diet and tonic medicines, outside and in. Although the restriction of the diet was very irksome and repugnant, still, with rare fortitude and firmness, she adhered strictly to the path marked out, and has been rewarded by a return to comparatively good health—she calls herself well— and the physical signs have almost entirely disappeared.

CASE IX. A woman with a very young child, that was especially devoted to crying by night, and keeping its mother awake, was very pale, thin, coughing severely, raising largely, with diminished resonance on percussion, and crackling throughout the upper third of right lung. This was in October, 1874. At the present time she coughs but very little; the physical signs are diminished; the softened lung tissue drying up. Her appearance has improved so much, that to an observer she appears well. All this time she has nursed her child. At present, owing to the hard times, she has been obliged to relinquish her meat diet; but through the benevolence of a lady I am able to have her keep up her diet a little while longer.

It is very interesting to see the effects of animal food in a complaint so universally fatal, and, unhappily, so very prevalent. More cases might be advanced from other physicians' practice, outside my own, tending to the same result. They show the great importance of diet in chronic complaints, and how that in tissue-wasting diseases a perfect pabulum should be afforded as a basis for proper treatment. It would be very desirable if some hospital, or sanitarium, for the treatment of consumption, could be established to test the effect of food upon a large scale. The cottage plan of hospital would be the only desirable one, as, in view of the present doctrine of the contagiousness of tubercle, one would shrink from crowding consumptives together in large wards. The *morale* of an organized hospital is more favorable for carrying out a regimen than the loose government of private families. The very idea of going to some specific place for a particular purpose gives a psychological element of firmness and decision of character not otherwise obtained. Here is a fine opportunity to devote one's posthumous wealth to the doing of a real good to mankind,—that is, to found a hospital for this purpose, and settle the question beyond a doubt.

The present general use of flour is condemned. If what has been said is true, there is no other course to be pursued than to condemn the idea and practice of trying to nourish and develop human tissues from flour. We advocate a return to wheat meal as a matter of national, dental, capillary, neurotic, medical, and physiological importance. What we want for the United States is a human race with good hair and teeth, good nerves and intellect, good eyes, and good vital tissues and fluids, that can fight the battle of life with disease, resist disease, and perpetuate a noble posterity of full physical vigor, excelling in strength and perfect in functions. As citizens, as lovers of mankind, and as lovers of God, we are bound to do our part towards this grand end. Given a sound mind in a sound body, what heights of glorious perfection may not be developed by education and culture; and, on the other hand, how much suffering and pain may be averted.

Liebig first propounded this doctrine, and it is opposed by Pavy, on the ground that we do get the proper modicum of salts in the mixed diet; but it is not certain that the diet of the average citizen in New England is "mixed" as much as he thinks. Potato is the ordinary staple vegetable food, which is combined with flour; but, according to Letheby, healthy potatoes contain seven tenths per cent. of saline matter; and, according to Payen, there is one and twenty-six hundredths per cent. of lime (pectates, citrates, phosphates, and silicates), magnesia, potash, and soda salts. On the other hand, as we have seen in potatoes that rot, the *Chemical News* has shown that four fifths of these lime salts are absent; so that Pavy may be wrong in his estimate, by overlooking the excessive withdrawal of salts from plants raised on exhausted soils.

Diet recommended. If flour is left out, or made to take the place it occupied in this vicinity in our diet fifty years ago, our present modes of living need not be much changed when people are in ordinary health. Animal food in this climate should predominate more than it does now. Some people need meat twice daily. Then there is the Indian-corn meal, the rye meal, the barley meal, beans, and pease. Food cooked by steam is preferable to boiled; because, in boiling, the soluble salts are dissolved out from the meat or vegetables, and a higher degree of heat is applied, because dry steam is hotter than boiling water. No more elegant, appetizing, or satisfying dish is ever served than cracked wheat, cooked by dry steam, and eaten with milk or cream.

The writer hopes to see the time when our cooking will be done by steam. He believes it possible, and should be glad to exhibit his devices for securing this object. Roasting and broiling do not remove the soluble salts, and hence are preferable to *boiling*. Broth is good food.

The preparation and cooking of food are ordinarily deemed

drudgery. That such an idea should prevail is one of the signs of a fallen nature. Cooking is a chemico-physical process, of great intricacy and refinement, for the purpose of rendering food digestible. It is accomplished by heat,— the wonderful laws of which are more abstruse than almost any other department of physics. Surely, it cannot be undignified, or unladylike, or unscientific, or unphilosophical, for the most refined and intelligent person to engage in pursuing labors in a laboratory, even it be called a kitchen ; or engage in producing remarkable results by the action of heat, even if it be called baking bread, or working over a cook stove. A more healthy public sentiment in relation to these things prevailed in the days of the patriarchs, when Sarah cooked the kid for Abraham and his heavenly guests.

In closing this discussion, we wish to appeal to the medical profession to occupy itself more in directing and moulding public opinion, by official utterances of their present organizations, in relation to diet, dress, and hygiene. The action of one individual alone can effect little or nothing. It was just so with the weather reports. For a long time there have been plenty of meteorological observers all over the country, but their labor, however painstaking and laborious, availed nothing of importance, until similar observers were combined together in systematic contemporaneous observations all over the country. And now we have weather predictions—eighty-five per cent. true. Suppose all the medical societies in New England should investigate the subject of flour, and should all come to conclusions which coincide with those laid down in this paper, and should officially announce the same to the public, would it be long before the directing and moulding influence would be felt generally, and for the benefit of the mass of mankind?

It is sincerely to be desired that the physicians will hasten these efforts, for it is high time that our diet list should be made out by physiologists, and not by some nameless

French cook; that our dress should be indicated and designed by intelligent artist anatomists, instead of some fashionable Parisian human being, who is ignorant both of the structure of the body and the functions of the organs contained in the great cavities; and, finally, that our hygienical conditions should be explained by medical physicists, instead of being left to whim, fancy, or even cupidity, as they now often are.

IS FLOUR

OUR PROPER FOOD?

WITH SOME REMARKS UPON THE EFFECT OF

ANIMAL FOOD IN CONSUMPTION,

BY

EPHRAIM CUTTER, A. M., M. D.,

OF CAMBRIDGE, MASS.

A LECTURE DELIVERED BEFORE THE N. H. STATE MEDICAL
SOCIETY, JUNE 16, 1875.

CONCORD, N. H.:
PRINTED BY THE REPUBLICAN PRESS ASSOCIATION.
1875.

www.ingramcontent.com/pod-product-compliance
Lightning Source LLC
Chambersburg PA
CBHW030710110426
42739CB00031B/1700